"Hello, I am a colour-changing chameleon. Look for me every time you turn the page!"
STRAWBERRY
LADYBIRD
TOADSTOOL
TOMATO
STOP SIGN
Who is driving the bus?
FIRE ENGINE

What's Orange?

How many legs does an octopus have?

Can you name an orange vegetable?
SQUIRREL
CARROT
TIGER
How many leaves does the carrot have?
BICYCLE

What's Yellow?

Which flowers did the bee land on?

What is the sun wearing?

What colours are the other bugs?

What's Green?

LIME
How many peas can you count?
PEAS
Which frog has caught the fly?
FROGS
LILY PADS

SNAKE
What colour are the snake's spots?
CABBAGE
TREE
DRAGONFLY
TURTLE
CROCODILE

What's Blue?

Which animal is the biggest?

What would you wear to keep yourself dry?

How many fish can you count?

What's Purple?

Can you spot the hidden worm?

Who is wearing the other spotted sock?

How many bees can you count?

What's Pink?

What colour are the candles on the cake?

WORM
CANDY FLOSS
BALLET SHOES
How many shells can you count?
NECKLACE
BLOSSOM
ICE CREAM
HIPPO
Which shell is a different colour?

What's Brown?

How many legs does the chair have?

What noise does a monkey make?

Can you spot two differences between the biscuits?

What's Grey?

Follow the stepping stones to help the baby elephant find its mum.

What noise does a wolf make?

Which animal has horns?

What's Black?

Can you get through the
the spiderweb maze?

How many legs does
a spider have?

What colours are
the bats' bellies?

What's White?

CLOUD
MOON
STARS
How many yellow stars can you count?
SWAN
POLAR BEAR
SNOW

What's MultiColoured?

What colours are the paints?

Which button is the odd one out?